MW00941336

# Cycle Count and Physical Inventory Design and Execution

## An Engineering Guide

### By Jan Young

# CYCLE COUNT AND PHYSICAL INVENTORY DESIGN AND EXECUTION

Copyright © 2010 by Jan B Young
All Rights Reserved

ISBN 978-0-557-36935-5

Cover images from Wikimedia Common. Cover design and all interior graphics by the author.

# Table of Contents

Introduction.............................................. 2
   About the Author ................................. 6
Cycle Counts........................................... 7
   Reasons For Cycle Counting .................... 8
   Error Correction ................................... 9
   Double-Checking ............................... 14
   Procedure Validation........................... 15
   Accuracy Measurement........................ 17
   How To Measure Accuracy .................... 18
   Inventory Accuracy Defined ................. 18
   Why Measurement Is Important........... 19
   How To Measure............................... 20
   Control Charts................................... 29
   Structuring A Cycle Counting Program..... 33
   How Much Cycle Counting Is Necessary?
   ...................................................... 34
   On-Line Cycle Counting Methods.......... 37
   Off-Line Cycle Counting Methods.......... 41
   Basic Principles ................................. 43
   Cycle Count Planning .......................... 45
   Recommendations for a Cycle Count
   Program.......................................... 47
Physical Inventories .................................. 51
   Planning A Physical Inventory................. 54
   Physical Inventory Methods.................... 58
Conclusions............................................ 65
Appendix – Tables .................................. 67

# Introduction

Publicly and privately owned companies are required to keep financial records in a relatively standard way so stockholders can read financial reports with reasonable assurance that profits are profits and losses are losses. Further, it is usually required that an independent auditing firm review the bookkeeping methods and testify annually to the appropriateness and fairness of the reports.

The value of inventories is a major item on the annual financial report of most companies. Any misstatement in inventory will directly affect profit. If an error overstates inventory, costs will be understated and profit overstated until the error is found and corrected. Then the correction will result in a reduction in the previously reported profit. If a similar error understates inventory, profit is understated at first, with the correction later resulting in an increase. Naturally, both internal and external auditors are vitally interested in inventory record accuracy.

Most inventory records are kept using the perpetual inventory system. Rather than recount inventories every time something changes, perpetual inventories are maintained

by recording only the changes and applying
them to the current on-hand balance on the
books. While perpetual inventory records are
inexpensive to maintain compared to the
alternatives, they are dependent upon the
accurate and reliable capture of all
transactions. When a transaction error
occurs, the perpetual inventory is thrown off
for at least one item and will not be corrected
until the inventory is recounted.

Unfortunately, the huge volume of
transactions involved, and the fact that semi-
skilled or unskilled hourly employees record
most transactions, makes keeping accurate
inventory records difficult. The idea of the
"annual physical" was originally an
accountant's way of assuring that inventory
investment records be as accurate as possible.
Over the years, those of us who are
responsible for inventory control and inventory
accuracy have proven the need for the
physical inventory process over and over
again.

The fact that paper driven inventory systems
are far from accurate has been known for a
long time. Over the years, a variety of things
have been tried to improve accuracy while still
keeping costs at a reasonable level. Two
methods are commonly used today: cycle
counting and the "full" physical inventory.

Cycle counting is a process by which portions
of the inventory are selected for counting on a
regular basis. People, called cycle counters,
are sent into the warehouse to physically view
and count the material actually on hand. The
results of these cycle counts are then
compared to the inventory records and used to
make corrections.

The full physical inventory is, in a sense, a
simultaneous cycle count of every location and
every item in inventory. The scope of the full
physical inventory is much greater than that
of a cycle counting program. And, for a variety
of reasons, the financial and auditing staff is
more likely to be involved in a full physical
inventory than in a cycle count. Therefore, the
systems and procedures used to support full
physical inventories are significantly different
from those used to support cycle counting.

The discipline of cycle counting has been
around for decades. It has the potential of
delivering major improvements in inventory
record accuracy while simultaneously paying
for itself in reduced overhead cost. Although
widely used in the largest and most
sophisticated warehouses, the actual payback
rate achieved in cycle counting programs has
varied significantly, in most cases because
management, supervisors and operators do

not understand the principles that underlie the concept. What is the purpose of cycle counting? How should it be done and how much of it is needed?

This engineering guide is intended as an introduction to both kinds of inventory counting for senior management and process engineers in the logistics, distribution and warehousing industries.

> It explains what can and what cannot be accomplished with a program of cycle counting. It describes how cycle counting should be structured and how it should fit into the organization. It reviews the math required to determine the optimum level of cycle counting and lays out a sample cycle counting program that includes the training required, both for the workers who actually perform the counts and for the supervisors who direct them and use the results.

> It discusses the reasons why full physical inventories may be necessary, the problems associated with accuracy, and the circumstances under which they can be eliminated. It reviews inventory-taking processes and discusses training and auditing

requirements and can serve as a guide
for the auditors and supervisors who
specify and control the process.

## *About the Author*

Jan Young is a trained Industrial Engineer
with thirty-eight years experience in
manufacturing and distribution. Now retired,
he has:

- o Managed a factory and a warehouse
  employing over a hundred workers in a
  24/7/365 continuous operation
- o Sold, installed, configured and
  maintained both manufacturing
  planning and warehouse management
  systems on a global scale for more than
  twenty-five years
- o Designed a warehouse management
  system. That system, aimed at the top-
  tier market, is now one of the premier,
  commercially-available systems
- o Consulted in more than fifty warehouses
  in ten countries

# Cycle Counts

Cycle counting, according to the American Production and Inventory Control Society, is:

> An inventory accuracy audit technique where inventory is counted on a cyclic schedule rather than once a year. For example, a cycle inventory count is usually taken on a regular, defined basis (often more frequently for high-value fast-moving items and less frequently for low-value or slow-moving items). Most effective cycle counting programs require the counting of a certain number of items every workday with each item counted at a prescribed frequency. The key purpose of cycle counting is to identify items in error, and trigger research, identification, and elimination of the cause of errors. (Wallace and Dougherty)

In practice, cycle counting often involves counting the product in specific locations and comparing quantities found with the records. Errors are usually brought to the attention of supervision and attempts are often made via historical records to locate the source of the errors. Management reports and summaries

can also serve to measure actual inventory
accuracy levels.

## Reasons For Cycle Counting

Cycle counting is an often-misunderstood
activity. Cycle counters are in a position
to achieve four separate goals as they work
their way through the inventory. These goals
are to:

1. Catch and correct inventory errors,
   thereby improving accuracy
2. Double-check the corrections that result
   from inventory errors found by others
3. Validate the proper use of procedures
   for recording and handling of inventory
   transactions and identify procedural
   problems
4. Measure the accuracy of the firm's
   inventory records

To put the cycle counting function in its
proper perspective and to show how an
effective cycle counting program can be
constructed, it is best to examine each of
these goals separately.

## *Error Correction*

Cycle counting, by itself, cannot turn inaccurate inventories into accurate ones. Cycle counters do, on occasion, find inventory record errors. They are in a position to correct the errors they find and should be charged with doing so. But the percentage of the errors that a reasonable level of cycle counting will correct is too small have a significant impact on the overall level of accuracy in most businesses.

To understand why this is so, consider an inventory system as containing its own "inventory" of errors. The figure on the right illustrates the inflow of errors made as material is moved through the plant and

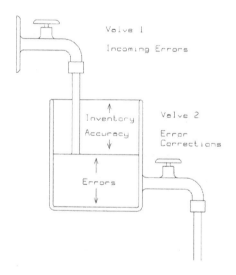

the outflow of errors made by cycle counters as they locate and correct them. Accurate inventories are achieved by keeping the level of "water" in the "container" low.

The rate of flow of errors into the container is
a function of the number of material handlers,
the rate at which they work, and the accuracy
of their work. Because some errors are posted
to already-incorrect inventories and should
not be counted, the rate at which errors flow
into the container is also a function of the
overall accuracy.

$$REI = NMH * RMH * RTE * IA \qquad \text{Eq. 1}$$

In this formula, REI is the rate at which errors
are inserted into the inventory, NMH is the
number of material handlers, RMH is the rate
at which material handler work, and RTE is
the rate at which material handlers make
mistakes. If, for instance, a business employs
50 material handlers, if each material handler
does 100 movements per hour, and if the
movements are performed and recorded with a
½ percent error rate, 25 errors per hour are
generated. If the inventory accuracy is 90
percent, then 10 percent of these errors don't
count and the net rate of new errors created is
22.5 per hour.

The rate of flow out of the container is a
function of the number of cycle counters
employed and the frequency with which they
can find and correct errors. Assuming that
errors are randomly distributed through the

inventory and that the cycle counters check locations selected at random, the frequency with which errors will be found and corrected is a function of the overall level of accuracy and the rate at which they work:

$$REC = NCC * RLC * (1 - IA)$$   Eq. 2

In this equation, REC is the rate with which errors are corrected, NCC is the number of cycle counters employed, RLC is the rate at which each cycle counter checks locations, and IA is inventory accuracy. So, if a business has four cycle counters, if they each check six locations an hour, and if 10 percent of the locations are incorrect (i.e., inventory accuracy is 90 percent), they will find an average of 2.4 errors per hour.

Over a period of time, assuming no outside influences, the overall inventory accuracy will stabilize. The rate of errors being inserted into the inventory will equal the rate at which they are discovered and corrected. We can, therefore, combine these two equations and solve for the resulting level of inventory accuracy.

$$\frac{(1 - IA)}{IA} = \frac{NMH * RMH * RTE}{NCC * RLC}$$   Eq. 3

A new term can now be defined: the ratio of the rate at which material handlers move material to the rate at which cycle counters check locations.  This term, R, is defined as:

$$R = \frac{NMH * RMH}{NCC * RLC}$$

Eq. 4

Substituting it into the equation for inventory accuracy:

$$\frac{(1 - IA)}{IA} = R * RTE$$

Eq. 5

Or

$$IA = \frac{1}{1 + (R * RTE)}$$

Eq. 6

So, inventory accuracy is a function of the amount of cycle counting done in comparison to the amount of material handling done and is a function of the accuracy with which movements are recorded.  This relationship (equation 6) is

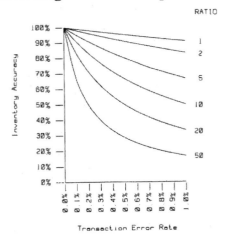

graphed on the right. It is important to remember that it represents only a steady state level of accuracy and that it assumes random cycle counting and a random distribution of errors in the inventory. Also, the equation and graph are a little optimistic because they assume that the cycle counters themselves do not make errors.

The two keys are the ratio of the speed of material handling to the speed of cycle counting and the accuracy with which transactions are recorded. If transaction accuracy is 99.5 percent, then to achieve a steady state of 75 percent inventory accuracy, Equation 6 implies that one cycle count must be performed for every sixty material movements. If the plant employs 50 full time material handlers and if a cycle count takes six times as long as a typical material movement, a staff of five full-time cycle counters will be required. And, incredibly, achievement of 99% inventory accuracy would require 380 cycle counters!

These high ratios are based on the important assumptions that cycle counters must search (either at random or in a planned fashion) to find errors and that there is no other mechanism in place for finding and correcting errors. Clearly, no sane cycle counting

program, by itself, can turn inaccurate
inventories into accurate ones.

On the other hand, the equations above can
also mean that, given enough time and labor,
inventory accuracy can result from cycle
counting by itself. Businesses that have a
small number of highly critical items can
assure accuracy for those items through
intensive cycle counting. If the number of
items is small and if the number of
movements made is also small, the cycle
counting labor required may be an acceptable
alternative to a complex and expensive
inventory system. At least one major
pharmaceutical company once tried this
approach on a large scale. They literally cycle
counted every storage location every time
material was moved into or out of it. The
problem then became one of finding a way to
check the checkers.

## *Double-Checking*

Material handlers will, from time to time, find
that the amount of material in inventory does
not appear to be what it should be. Some of
these exceptions will, in fact, be instances in
which the records are incorrect. Others will
be instances in which the records are correct,
and the material handler is wrong. As
inventory accuracy increases, the second

cause becomes more important. The number of errors put into the inventory by erroneous corrections then becomes a significant concern.

Requiring that a second person confirm the need before applying a correction to the inventory database can provide some protection against erroneous corrections. For instance, when a material handler finds an error such as a shortage or material in a location that is supposed to be empty, he or she should be required to determine and enter the correct on hand quantity on the spot, but the correction should be posted to the database only after a cycle counter has visited the location and agreed with the material handler's adjustment. The protection can be extended one step further if the system is designed so that only cycle counters can make inventory adjustments. Even the senior manager, then, must enlist the aid of a cycle counter to post an adjustment. By creating people who are specialists in verifying and correcting inventory balances, fewer errors are created and overall inventory accuracy levels are enhanced.

## *Procedure Validation*

From time to time in even the most accurate inventories, the people doing cycle counts will encounter real errors. If the inventory system keeps adequate historical records of the movement of material, it should be possible to track down the reason why the error occurred. Once the source of the problem has been located, management has an opportunity to prevent future occurrences.

Although the validation of procedures is a reasonable goal for cycle counting, it is a difficult one, even with the best audit trails. If, for instance, an inventory is incorrect because an unrecorded withdrawal was made, how is the cycle counter to trace the lack of a transaction to its source?  Or if, for instance, a quantity was incorrectly picked, how can we take action to prevent recurrence if the error was random and the person who made it is otherwise a valuable employee?

Further, the tracing of errors through history files and other records requires a great deal of time and effort. Often the people who physically perform cycle counts are trained only for the counting process and the tracing job falls to salaried people.  When many transactions affect an inventory over a short period of time, the workload can be enormous.

Still, at least to some degree, cycle counting can help ensure that proper procedures are

being followed and that the best possible job is
being done to keep the inventories accurate.

## *Accuracy Measurement*

Knowledge of a firm's actual level of inventory
accuracy is important. It serves as a
management guide and provides the
accountants and auditors with assurance that
the books properly reflect the actual value of
the company's investment in inventory.
Representations to the stockholders, the
government, and management must be
accurate.

Inventory accuracy is also important to both
the inventory manager and the warehouse
manager because inaccuracy can cause
material shortages, unneeded expediting,
production delays, customer service problems
and wasted labor. Inaccuracy can also result
in a decision to increase safety stocks,
consuming both valuable storage space and
investment capital.

Operating without a measure of the accuracy
of the inventory records is like flying a plane
without instruments or driving country roads
without a map. The results are hit and miss.
Inefficiency, poor customer service, and, in a
competitive environment, even business
failure can result.

# How To Measure Accuracy

## Inventory Accuracy Defined

Before inventory accuracy can be measured, it has to be defined. For the purposes of this book, and for the purposes of inventory and warehouse managers and systems, an inventory record is accurate if the corresponding storage location contains all of the correct part numbers and if a physical count of the parts agrees precisely with the recorded quantities. In addition, if material is accounted for by job, the recorded job numbers must be correct and the quantities allocated to each job must total to the actual quantity on hand. Similarly, when lot or serial numbers are tracked, they must be correct. If any single element of data is not in agreement with the record, the record is incorrect. Inventory accuracy is the percentage of locations for which the records are correct by the above definition.

Accuracy on a location-by-location basis is important because material handlers are required to find the material when told what location it is in. If, in a large inventory facility, material is stored somewhere other than the recorded location, the effort required to find it can be so great that the material might just as well not exist. Four-wall inventories are useful

to the bookkeeping department; they are
nearly worthless to the people who actually
handle the stock.

An accuracy tolerance can be allowed when
counting the content of a storage location that
contains a bulk material such as a fluid or a
powder. These materials sometimes stick to
the sides of their containers, and may be
subject to evaporation and other losses.
Further, it is often impossible or impractical to
measure quantities to enough significant
places to prevent little errors from
accumulating into larger ones.

In some instances, an accuracy tolerance can
also be allowed when dealing with low-cost,
physically small items like fasteners. As with
fluids, it may be impractical to count these
items precisely, even with the most accurate
electronic scales. However, for most items, the
definition of inventory accuracy assumes a
zero tolerance for counting errors and
identification errors.

## *Why Measurement Is Important*

Surprisingly few companies actually measure
the accuracy of their inventories. If inventory
accuracy is stable and the records are known
to remain accurate, measurement can be
infrequent and, therefore, inexpensive. If,

however, recordkeeping methods change often or if personnel turnover is high, more frequent measurement is needed. Management needs to have a handle on the level of record accuracy so that it can take action when the level is unacceptable or when a downward trend becomes evident. Without measures, management is blind, assumptions are made, and problems are ignored until they grow to the point where they do real harm.

## *How To Measure*

The annual physical inventory is typically not a valid measure of accuracy for two reasons. First, as discussed on page 51, the annual physical inventory itself is almost always inaccurate and can introduce more errors into the inventory records than it removes. The result is that it represents a biased and inaccurate measurement. And second, physical inventories are usually done too infrequently to serve as a reasonable measure of accuracy. Good inventory management demands a fresh measure of accuracy at least monthly and probably more often than that.

Many companies no longer do physical inventories because they have proven to themselves and to their auditors that physicals are not needed. For these companies, the annual physical inventory is

simply not available as a measure of accuracy.
Cycle counting, when done by well trained
employees in a "do-it-right" atmosphere, may
be management's only real handle on
inventory accuracy.

Basic Measurement Techniques

Cycle counting programs are closely related to
quality inspection programs that draw
samples from the parts on, for instance, a
pallet. The quality of the parts in the sample
is measured and is used as an indication of
the quality of the entire group. Similarly, cycle
counters can sample the inventory and use
the accuracy of the sample as an indicator of
overall accuracy.

The techniques required to measure inventory
accuracy can, therefore, be adapted from ones
used by quality control. QC, after all, has
been measuring the accuracy of the
production workforce for years. The only
extension is to use the techniques to measure
the accuracy of the material handlers.

Inventory accuracy should be measured by
randomly selecting a number of storage
locations and sending a cycle counter to verify
that these locations contain exactly what the
inventory control system says they contain. It
is important to note that random selection of

the locations to be cycle counted is important. Statistical sampling theory requires this randomness. None of the methods and techniques presented here are valid if they are applied to a non-random group of locations.

The number of locations to be cycle counted each time period (day, week, or month) is shown in table 1 on page 67. Very precise measurements require huge samples. But reasonably accurate measurements require only modest samples. If accuracy is believed to be 70 percent, and needs to be confirmed within plus or minus 10 percent, the required sample size is only 189 counts. If accuracy is 98 percent and needs to be confirmed within 10 percent (i.e., that accuracy is greater than 88 percent), then only 18 counts are required.

The degree of precision defined in the table accounts for variations in measurement because the process is based on sampling techniques. The sample chosen, being random, may not truly represent the overall accuracy. The degree of precision does not, unfortunately, account for errors made by the people doing the cycle counting. Sampling theory assumes that inspectors and cycle counters do not make mistakes. So, they must be well trained and motivated to work carefully. Any excess emphasis on speed in cycle counting will tend to reduce its precision in unpredictable ways.

Inventory stratification, also known as "ABC analysis," may affect the way inventory accuracy is measured, since accuracy goals can properly be a function of inventory class. Suppose a business is maintaining 99.5 percent accuracy for A items, 95 percent accuracy for B items, and 90 percent accuracy for C items. If the measurements are to reflect actual accuracy within plus or minus 2 percent, then a random sample of 112 A item locations, 1069 B item locations, and 2025 C item locations must be drawn and those locations must be counted. It might also be decided that the measurements for B and C items need be less accurate than those for A items, limiting the total number of counts required to obtain a measurement. Clearly, as the number of levels of stratification increases, the amount of counting required goes up and the cost of measuring goes up.

Some people are surprised to find that the sample size required to measure accuracy is not a function of the number of parts or locations in an inventory class. This is because the technique assumes a homogeneous population of inventory records and because samples are drawn randomly. The amount of salt in seawater can be

determined as easily from a cupful as it can
from a tank car full.

The count requirements table (page 67) is
calculated from equation 8 below, which may
be used in its place if, for instance, the
number of locations to be counted is to be
determined by a computer system.

The number of locations to be cycle counted
each time period depends on the number of
independent measurements management
wants during that period. And, for each
measurement, it depends on the precision
with which management wants accuracy
measured. Suppose, for example,
management decides that a fresh
measurement of inventory accuracy is needed
for each of three product lines each week. This
means that three samples must be drawn
weekly, each random but limited to locations
containing parts for one of the three product
lines. During the week, these locations will be
counted and the counts compared to the
inventory records. At the end of the week, the
number of counts made for each product line
will be divided into the number of locations
found to be in error in each sample. The result
will be an estimate of the inventory accuracy
for that product line:

$$Inventory\ Accuracy = \frac{Number\ of\ Errors\ Found}{Number\ of\ Counts\ Made}$$

Eq. 7

If more than one error is found in a location, the location is considered to be in error and the two errors count only once[1].

The number of locations to be counted in each of the samples should be a function of two things: the expected level of inventory accuracy and the desired counting precision. Well-established sampling theory provides the following equation:

$$NC = \frac{EIA*(1-EIA)}{(P/3)^2}$$

Eq. 8

NC, the number of counts required in a sample, is calculated from the expected inventory accuracy (EIA) and the precision required (P). Both are expressed as decimals (i.e., 3 percent is 0.03) for calculation purposes. It is interesting to note that the more accurate inventories are, the fewer counts it takes to prove them accurate. And, the number of counts required to achieve a

---

[1] But if the location contains more than one item and the errors relate to different items, they should be counted separately.

measurement goes up with the square of the required precision.

So, for example, imagine a business that feels that it needs a fresh measure of inventory accuracy each month. If that business feels that it has 95 percent accuracy (maybe demonstrated by an earlier measurement), and if it requires that measurements be accurate within plus or minus 2 percent, the number of counts required is:

$$NC = \frac{.95*(1-.95)}{(.02/3)^2} = 1069$$

Thus 1069 counts must be performed each month. If the 1069 counts are performed and 50 errors are found, then the overall accuracy may be estimated at something between 93.3 and 97.3 percent, slightly better than the first estimate. (The calculations are as follows: 50 errors divided by 1069 locations sampled is a 4.7 percent error rate, or a 95.3 percent accuracy rate. Given that the measurement is plus or minus 2 percent, accuracy is known to be between 97.3 and 93.3 percent.)

And, once again, it is important to understand that the equation will not produce valid results unless random locations are counted. Locations that are cycle counted for other

reasons (such as the discovery of an error or
the scheduling of an item for routine periodic
cycle counting) should not be included in
accuracy measures because they introduce
biases. Other circumstances may limit the
usefulness of the equation and expert
statistical help may be advisable before relying
on it heavily.

Recalculating Confidence

Inventory accuracy sample sizes are calculated
based on an expected level of accuracy.
Initially, the expected level of accuracy is
probably a guess, made by someone
knowledgeable. After some experience has
been gained, expected levels should be
reasonably well known. However, there can
be times when the expectations are not borne
out by the results of the sample.

Whenever locations are cycle counted and
inventory accuracy is calculated, the results
should be compared with the value of EIA
(expected inventory accuracy) used to
determine the sample size. If the first estimate
differs from the value of EIA by more than a
small amount, the formula used to select the
sample size should be worked backward to
calculate the actual tolerance level achieved.
In other words, having obtained an estimate of
the inventory accuracy, the actual precision

available from this estimate should be
calculated from:

$$P = 3 * \sqrt{\frac{IA*(1-IA)}{NC}}$$
Eq. 9

In this equation, P is the precision, as defined
previously, IA is the actual inventory accuracy
that resulted from the sampling, and NC is the
number of counts performed.

If, for instance, a company estimates its
inventory accuracy at 95 percent and wants to
sample within plus or minus 2 percent, the
proper sample size is 1069 as calculated
previously. If, however, the cycle counters
find 125 errors in the 1069, the result of their
work is an accuracy estimate of 88.3 percent.
Then, the precision of the 88.3 percent
estimate can be calculated as:

$$P = 3 * \sqrt{\frac{.883*(1-.883)}{1069}} = 0.0295 \ or \ 2.95\%$$

This means that the company knows for sure
only that its inventory accuracy is somewhere
between 85.35 and 91.25 percent. If a
tolerance level of 2 percent is important, then
a new sample size must be calculated using
equation 8 with an estimated accuracy of 88.3
percent. The results of this calculation will

require that additional locations be randomly
selected and cycle counted. Following the
completion of these cycle counts, it will be
possible to calculate a new estimate that, if
close to 88.3 percent, will be accurate within
the original 2 percent tolerance.

## *Control Charts*

Now that a sample has been drawn and a
measurement obtained, the next step is to
decide what to do with it.  One can, of course,
make immediate judgments about the quality
of the inventory records, but it may be even
more useful to look at a series of
measurements over time. A statistical
technique known as the control chart is
valuable.

The concept of a control chart is simple.  Its
purpose is to detect changes in an ongoing,
stable situation. The chart is simply a graph
with inventory accuracy on the vertical axis
and time periods on the horizontal axis.  A
horizontal line across the center of the chart
defines the desired or expected accuracy.  A
second line, called the lower control limit
(LCL), defines the point at which action should
be taken.

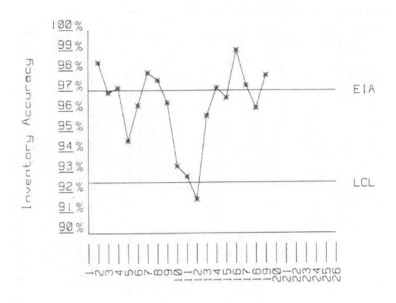

The figure above shows a sample control chart
for a hypothetical company that has
historically been achieving 97 percent
inventory accuracy. This company measures
accuracy weekly to within plus or minus 2
percentage points by doing 655 counts each
week as specified in the count requirements
table on page 67. Management often examines
the control chart looking for either of two
things:

1. One or more samples below the lower
   control limit of 92.3 per cent. If such a

sample exists, there is a likelihood that
inventory accuracy has actually slipped.

2. A consistent downward trend. Obvious
   trends indicate the possibility that
   inventory accuracy is currently slipping.

The chart in Figure shows a downward trend
beginning in week 8 and culminating in week
12. The measurement plotted for week 12 is
below the lower control limit, indicating a
problem. Judging from the later
measurements, management found the
problem and fixed it.

Control charts are easy to construct. The only
technical point is the placement of the lower
control limit, which may be calculated from
Tables 2 and 3 (pages 68 and 72). Begin by
selecting a value from Table 2 based on the
expected inventory accuracy and the size of
the samples used to measure accuracy. Then
select a value from Table 3 and multiply the
two values together. The result is the space
between the inventory accuracy line and the
lower control limit line on the control chart,
stated in percentage points.

The control limit confidence factor in Table 3
is derived from the statistical notion of a
normal curve. In essence, to determine a
value, the user of a control chart must first

decide how much time should be spent on wild goose chases. Because samples are randomly chosen, there is always the possibility that results aren't representative of real life. Therefore, it is necessary to consider the possibility that an unusually low measurement is, in fact, only normal day-to-day variation in the measurements themselves.

If, for instance, a person wishes to be 90 percent sure that a low measurement truly represents a problem before he or she spends the time to investigate, the factor value used should be approximately 1.3. The hypothetical company whose control chart is illustrated above chose a factor of 0.7 from the table, which gives them a 25 percent chance of chasing a wild goose but also gives them a lower probability of missing a real problem.

Those who wish to write software to prepare control charts need to know the underlying mathematics. The lower control limit is spaced from the expected value of accuracy according to the formula below:

$$SCL = F * \sqrt{\frac{EIA*(1-EIA)}{N}} \qquad \text{Eq. 10}$$

Where SCL is the control limit spacing, F is the control limit confidence factor, EIA is the

expected inventory accuracy, and N is the
sample size. All terms to the right of the
equals sign excepting F are included in Table
2. The calculation of values for F is complex
and beyond the scope of this book, so Table 3
should be stored in the inventory system and
used directly.

## *Structuring A Cycle Counting Program*

Businesses sometimes decide to cycle count
everything once a year and the A-items
once each quarter. Often, this is done with
little or no reason - it just sounds good.
Others worry when they are unable to afford
more than one or two people for cycle
counting, believing that more cycle counters
would provide more value.

Actually, the right amount of cycle counting is
much more reasonable.
Cycle counting is most effective when the
program is steady and regular, but modest. If
inventory accuracy is relatively good, it does
not take much work to prove that accuracy
has remained good, and few errors will be
found that need to be investigated. If
accuracy is poor, it doesn't take much work to
prove that either.

## How Much Cycle Counting Is Necessary?

There are costs and benefits associated with cycle counting. In theory, cycle counting should be done to the point where the last cycle count just barely pays for itself; an EOQ-like concept. In practice, however, let's think in terms of a different management decision process.

Answer two questions:

1. How often do we need to have inventory accuracy measures: weekly, monthly, or bi-monthly? How often and how quickly does inventory accuracy change? When a change occurs, how quickly must corrective action be taken?

2. How precisely do we need to know the level of inventory accuracy? If accuracy is 93% and as reported as 95%, is a problem created? A tolerance of 2% to 4% is often adequate.

Starting from your answers to these two questions, equation 8 (repeated below) gives the number of counts needed, NC, based on your known or assumed inventory accuracy (EIA) and the desired precision (P).

$$NC = \frac{EIA * (1 - EIA)}{(P/3)^2} \qquad \text{Eq. 8}$$

The number of counts required is the number required every time management wants a fresh measure of inventory accuracy. If, therefore, measures are required weekly, NC is the number of counts required each week.

In addition to measuring inventory accuracy, a good cycle counting program should also be used to validate procedures by identifying errors, tracking down underlying causes, and thus allowing corrections to be made. Unfortunately, the amount of cycle counting required to locate and correct errors in methods and procedures is not as easily calculated. To some degree, lower levels of accuracy imply added methods problems and therefore suggest the need for higher levels of cycle counting to identify the causes of error and allow corrective action to be taken. In fact, the amount of cycle counting required to validate procedures is almost entirely a matter of judgment. But, we can supply a little structure to the decision.

Possibly the best approach to cycle counting for procedure validation purposes is to cycle count locations where an error is found by material handlers in the course of their normal day-to-day work.  If material handlers are charged with responsibility for reporting inventory errors whenever they are found, then cycle counters can be dispatched to verify the correct inventory and to do whatever investigation is required.

Despite the concerns mentioned in the first few pages of this guide about using cycle counters to correct inventory inaccuracies, this approach has the side benefit of correcting at least a small number of errors, while also validating and critiquing procedures.  But, even this will not result in a dramatic improvement in inventory accuracy because material handlers are likely to find errors only when a shortage results.  Up-side errors (in which there is physically more inventory than recorded) will probably not be noticed.  Likewise, downside errors will probably not be noticed unless the result is zero on-hand.

If we accept that the discovery of an error by a material handler should result in a cycle count, then one estimate of the right amount of cycle counting for procedure validation is the number of errors found.  Since few up-side errors will be found, the estimate might be reduced to the number of shortages occurring in a normal period.  One could also decide that the cycle counters should only investigate a sample of the errors. Such a decision would further reduce the cycle counting effort required.

How much is enough?

How much cycle counting is enough? The sum of the number of cycle counts required for measurement and the number expected to be required for procedure validation is enough.

Actually, cycle counting is most effective when the program is steady and regular, but modest. Modest cycle counting programs, therefore, can achieve most of the results of massive programs, and can do so with much greater efficiency and at much lower cost.

## _On-Line Cycle Counting Methods_

The basic cycle counting method is simple: employees are sent into the warehouse with instructions to visit pre-selected locations, identify and count the material found in them, and report the results back for comparison with the inventory records. The results are then used to measure accuracy, maintain audit trails, and make adjustments to the records.

For example, in an on-line environment, a
cycle counter might carry a hand held radio
terminal. The terminal begins the cycle
counting routine by specifying the location to
be counted. The cycle counter walks to that
location. Since there is a possibility that he or
she might select the wrong location, the
terminal requires that a location label bar code
be scanned.

Next, in an environment where product is
stored in unit loads, the cycle counter is
prompted to scan the bar code on one of the
unit load labels in the location and then
identify and key (or scan) the part number.
The system verifies that the unit load does
belong in the location counted and that it does
contain that part number. Finally the counter
is required to count the number of pieces of
that particular part number on that unit load
and the system verifies the quantity. If
everything checks out, the system prompts for
the next unit load label[2]. When the counter
indicates that the location is complete, the
system verifies that its records show no

---

[2] Count effort can be significantly reduced without
materially affecting accuracy if the counter is allowed to
specify that the pallet is "full." The system can then
look up the full pallet quantity for the item and compare
that number with the recorded inventory. This short
cut works best when the difference between a full pallet
and a partial one is easily discernable.

uncounted product in the location and then
displays the next location to be counted.

Exceptions will be found from time to time.
Unit loads will be found that are not supposed
to be there and some that should be present
will be missing. Items will be found that do
not belong on the unit load and other items
that should be there will not be. And
quantities will be incorrect.

When unit load numbers are found to be
incorrect, the system should revalidate the
location number and then search for the
unexpected unit load in its database. If the
unit load is recorded as being elsewhere, the
system should add it to the location being
counted, remove it from its recorded location,
and put that location on the list for cycle
counting. Serious errors such as this should
also be brought to management's attention for
further investigation.

When item numbers are found to be incorrect,
the system should revalidate the location
number and display descriptions of the item
number found by the cycle counter together
with the descriptions and item numbers of the
items that are on file as being in the location.
The cycle counter should then be allowed to
confirm the item number or change it. If the
item number is confirmed, the system should

record the existence of that item in the
location being counted and put all other
locations containing the item on the list for
cycle counting.

When quantities are found to be incorrect, the
system should prompt for a recount. If the
cycle counter confirms his or her count, an
adjustment should be posted to make the
recorded quantity match the quantity actually
found. Or, if the recount results in a number
that matches neither the recorded on hand
balance nor the first count, several actions are
possible. Some systems will ask for another
re-count, some will plan the location for
counting by another cycle counter, and some
will simply accept the second count,
regardless.

Those inventory facilities that do not use or
number unit loads can modify the cycle
counting methods described above by simply
eliminating input and verification of unit load
numbers. When lot numbers and/or serial
numbers are tracked as product moves
through the warehouse, they should also be
verified during the cycle counting of a location.

In some businesses it may be difficult for the
cycle counter to determine the item numbers
for the products found in a location. Many
distributors, for instance, handle products

that are clearly marked with the vendor's item number but do not bear the company's own item number. These businesses will often compromise by displaying the item number and description to the cycle counter and asking him or her to verify the number rather than requiring that it be independently determined. Some companies will even go so far as to assume that item numbers are correct and verify only quantities. Their argument, often, is that if the item is wrong, the quantity will almost certainly also be wrong.

## *Off-Line Cycle Counting Methods*

Cycle counting in an off-line environment involves some significant complications relating to the timing of the cycle counts and other activities that are going on in the warehouse. Imagine, for instance, an inventory facility in which the various tasks required are printed on worksheets and distributed to employees to be done. The employees check off the work as it is done and return completed worksheets to a clerk for key entry. When a cycle counter verifies the content of a location, there is no way for the system to know which of the outstanding picks and put-aways were complete and which were still pending at the time the count was done. Therefore, it has no way of knowing

whether the quantities found by the cycle
counter are right or wrong.

Cycle counting in an off-line environment can
only be done by freezing locations, counting
them, and then releasing them. Freezing is
usually done by setting a flag on the location
record to prevent new transactions from being
generated for that location. Once the existing
open transactions have been completed and
confirmed back to the system, the cycle count
can be done and the results entered. Then
and only then can the system be allowed to
create new material movement transactions
for the locations that were counted.

Some companies with off-line inventory
systems create a priority scheme that the
system uses to determine the relative
importance of transactions. When, for
instance, material arrives in receiving and is to
be put away, the system might recognize the
put-away transaction as less critical than the
cycle count. It would, therefore, hold the put-
away transaction until the cycle count was
complete. On the other hand, if a customer
order requires picking from a location to be
cycle counted, the cycle count might be
canceled to allow the pick to take place
immediately.

These complications apply just as much to
locations that are partly serviced by on-line
and partly by off-line material handlers. They
are one of the more important justifications for
on-line, radio-based inventory systems.

## *Basic Principles*

Regardless of the situation, several basic
principles are important to the design of an
effective cycle counting system. To be effective,
cycle counting employees must count
locations, not items. It is far from simple for a
counter to reliably locate all of the places
where an item can be stored. Finding and
counting all of the items in a location, on the
other hand, is easy. However, when an item
(or a specific lot of an item) must be counted,
the inventory system can determine which
locations are involved and the cycle counter
can count all of those locations. Don't forget
that cycle counting accuracy is as important
as transaction accuracy.

The use of the computer to select the locations
to be counted removes normal human biases
and allows more sophisticated item selection
methods. Methods used to select locations for
counting are described in more detail in the
next section.

Whenever a location is picked to a zero
balance, some businesses with on-line
inventory systems ask their material handlers
to verify that no material remains. This is a
very cheap and accurate way of getting one
more cycle count done.

Audit trails should consist of two parts. The
date, time, and transaction number of the last
cycle count and of the last material handling
transaction should be stored on the location
record and updated as work is done. In a
separate audit or history file, every material
movement and every cycle count should be
recorded. These records should specify the
location affected, the date and time the work
was done, the identity of the person who did
the work, the type of work done, and the
before and after on-hand balances.

Some businesses store the identity of the
person who did the last material movement on
the location record. When locations are
assigned to people for counting, this allows
them to assign someone else to do the count,
eliminating, or at least minimizing, the
temptation to cheat.

Cycle counting can be done either by
dedicated cycle counters or by existing
material handlers as part of their normal
workload. Using existing material handlers

saves labor (travel time into and out of the
stockroom or warehouse). More important, the
impact of being required to clean one's own
house will motivate better performance in the
recording and performing of transactions.
Using dedicated cycle counters, on the other
hand, allows for more detailed training and
reduces the likelihood that cycle counts will
suffer when the workload increases elsewhere.
Productivity and error rate measures should
be provided for cycle counters as well as for
material handlers. Many warehouses blend
these two extremes by selecting several senior
material handlers and giving them additional
training so they can effectively cycle count
part time.

## _Cycle Count Planning_

When a material handler finds an inventory
error, good practice specifies that the location
should be frozen and a cycle counter should
be brought in to verify that an error exists and
make the actual correction. Many businesses
prefer to make these kinds of cycle counts on
a priority basis to assure that working
locations are tied up as little as possible. Part
of the cycle counter's day, therefore, is spent
reacting to problems found by the material
handler.

The majority of the cycle counter's time,
however, should be spent counting in a
planned and orderly fashion.  To achieve this,
each day's cycle counting should be preceded
by a planning process.  In the planning
process, a supervisor or planner uses the
inventory system to determine the amount of
cycle counting that should be done and to
select the locations to be counted.  The
planning process results in a work queue of
locations to be counted.  The work queue can
be either transmitted to the cycle counters via
radio or can be printed and handed out in the
form of a worksheet.

The cycle count planning process typically
begins with a supervisor or planner entering
the number of cycle counters who will be
working and the number of counts that should
be generated for each.  Based on this
information, the system will select the
locations to be counted and will display them
for approval.

The locations to be cycle counted should be
selected using a combination of several
methods, with the number selected by each
method either predetermined and stored in a
system table or specified by the supervisor or
planner.  The selection methods are:

1. Randomly selected locations. The selection should be truly random, without bias for inventory value, ABC code, or the number of pieces on hand.
2. Locations that have gone the longest since last being cycle counted.
3. Locations that have had the most activity since last being cycle counted.
4. Locations specified by a supervisor or by the planner.

As the day proceeds, this predefined work queue will usually be supplemented with high priority transactions generated by material handlers as they discover count errors that require validation during their normal work.

## *Recommendations for a Cycle Count Program*

Although the proper role of cycle counting is achieved when its objectives are economically met, there remain open questions about the operation of such a program. Based on experience gained while consulting with a number of companies and in the line management of several shops, here are nine recommendations for the creation and operation of a sane cycle counting program. It is important to note that these recommendations are based on "typical" practices. You, the reader, must decide for yourself whether or not they are appropriate in the context of your business.

A. Do everything possible to assure that inventory transactions are accurately recorded. Validate the incoming data against all other possible data. Use bar coding to eliminate transposition of digits, incorrect item identification, and other human error. The more accurate your inventory records are, the more effective (and less expensive) your cycle counting program will be.

B. Use existing material handling people for cycle counting, rather than dedicated, "professional" cycle counters. Although this article has discussed "cycle counters" and "material handlers" as though they were different people, we were actually discussing only roles. In fact, it is best to use the same personnel for both jobs. Combining the jobs saves labor (travel time into and out of the stockroom or warehouse). More important, the impact of being required to clean one's own house will motivate better performance in the recording and performing of transactions.

C. Count by location, not by item. Don't forget that cycle counting accuracy is as

important as transaction accuracy.
Counting by location is much easier,
faster, and more accurate than by item.

D. Use computer assignment of the
locations to be counted randomly. This
removes bias and allows more
sophisticated item selection methods.

E. Track the date, time, and transaction
number of the last transaction on each
location and record who performed that
transaction. When locations are
assigned to people for counting, assign
someone other than the last person who
touched that location. This eliminates,
or at least minimizes the temptation to
cheat.

F. Limit the power to enter inventory
adjustments to cycle counters, as
authorized by a supervisor. Don't let
anyone (even the Manager) enter an
adjustment without first cycle counting
the location. This makes the material
handlers/cycle counters fully
responsible for the results of their work
and keeps "amateurs" out of the way.

G. On a day-to-day basis, the first priority
for cycle counting should be detected
errors; the second should be a specific

number of randomly selected locations.
The selection of locations should be
truly random, without bias for inventory
value, ABC code, or the number of
pieces on hand. A third priority, which
is optional, is to cycle count all or a
fraction of all locations that are picked
to zero inventory each day. The
justification for doing this is simplicity
and speed (nearly anyone can count to
zero effectively).

H. Expect your total cycle counting
program to consume between 1% and
3% of your overall material handling
labor.

I. Install proper controls over your cycle
counters and the cycle counting
program itself. In particular, the
following are important:

a. Limitations on who can post
inventory adjustments,
b. A complete inventory transaction
audit trail,
c. A complete cycle count and
adjustment audit trail, and
d. Productivity and error-rate
measures for both material
handlers and cycle counters.

# Physical Inventories

The original idea behind the physical
inventory was financial. Its primary goal
was (and still is) to provide a fair and unbiased
count of the dollars invested in inventory so
the company can give stockholders an
accurate picture of profitability.

The advent of computer-run inventory systems
created a second use for the physical
inventory -- that of recording the quantity on
hand for each part to verify the inventory
records. But, since the original concept was
designed only to provide an accurate overall
total, the second use has not worked well.
Most physical inventories result in error rates
between 5 and 20 percent by location. So long
as these errors are unbiased -- that is, so long
as they are equally divided between too high
and too low -- the resulting inventory is
accurate enough for accounting purposes.
But, there are obvious consequences for
material planning.

Problems inherent in most physicals include:

1. Their temporary nature. An entire
   organization is often built for physical
   inventory taking, complete with titles
   and lines of authority. But, since

everyone in the organization knows that it is temporary, there is little loyalty or dedication.

2. Pressures to get done and get back to work. The people doing the physical inventory know that the main business of the company is not to count inventory. They are often anxious to get back to their normal jobs. Pressure from management and customers to keep the plant on schedule can result in sloppy work and inaccurate results.

3. Lack of training. The people who do the counting and recording of physical inventories are usually taught how they should do their jobs but are rarely taught why. As a result, they do not often appreciate the consequences of failure and the importance of the physical inventory in total.

4. Inadequate performance measurements. In many cases, the only measure of performance that emerges from a completed physical inventory is the financial one: the size of the inventory write-down. Sometimes location accuracies are also measured. But, if individual inventory takers were measured on the speed and accuracy

with which they worked, there would be a significant improvement in the amount and quality of work done.

5. Inadequate auditing. Financial auditors who are interested only in dollars almost always do the auditing of physical inventories. Rarely do personnel with concern for individual location accuracies audit these inventories.

The net result is that physical inventories often insert more errors into the inventory records than they remove. The most successful companies no longer do physical inventories. Instead, they invest in high quality processes and procedures and in extensive employee training so that inventory records are accurate. They then use cycle counters to demonstrate accuracy. Auditing firms are almost always willing to accept existing records based on an audit of the effectiveness of a company's cycle counting program.

Some years ago a firm in the upper Midwest installed the systems they needed to achieve highly accurate inventory records. Later, the company's parent corporation found itself in bankruptcy, complete with indictments against the parent's management for fraudulent inventory records. One day, the

subsidiary company was paid a surprise visit
by three teams of auditors representing the
court, the parent company's board, and the
original auditing firm. The visitors were
prepared to close the company down to take a
complete inventory, but they chose to look
first at a sample of the company's cycle
counting records. When they found that the
company was keeping records that were
significantly more than 99 percent accurate,
they agreed to accept the inventories as
recorded and, within a matter of hours, were
gone.

## *Planning A Physical Inventory*

A well-run physical inventory meets two
sets of requirements: those of the
company's auditors and those of the inventory
control organization. The requirements
imposed by auditors can be hard to predict in
a paper such as this one because auditors
vary in expertise and experience. However,
the basic concerns involved should include (1)
assurance that the business' assets are
properly represented in the financial ledger, (2)
assurance that management information
generated by the inventory system is accurate,
and (3) detection of any possible instances of
theft or fraud.

The inventory control and warehouse operations organizations need precise knowledge of the part number, quantity, lot number, and serial number on hand at each storage location in the warehouse. They need this information to support high levels of customer service and operating efficiency at the lowest possible level of inventory investment.

Physical inventories are expensive exercises that are not always necessary and need not always be done annually. Before a physical inventory is started, a conscious decision should be made based on the costs and benefits involved. First, the cost of running the physical inventory should be estimated including both labor cost and the opportunity cost of being essentially out of business while the count is being done. Then inventory accuracy should be measured (see page 18).

With this information in hand, two informed decisions can be made by management and the auditing staff: (1) Is a physical inventory really necessary and (2) must it cover the entire operation or can selected areas be isolated for counting?

The physical inventory requires more than software. Organization, training, and supervision are also important parts of the

process.  The planning process should begin
several months in advance of the scheduled
inventory date with the creation of a cross-
functional committee.  The committee should
usually be made up of representatives from
finance, manufacturing, warehousing,
inventory control, purchasing, data
processing, and auditing.  It should be
charged with complete responsibility for
planning and executing the physical inventory.

The committee's work is divisible into tasks,
described here in approximately chronological
order:

1. One of the committee's first jobs should
   be to create a list of the tasks that must
   be done and the events that will occur.
   This list will undoubtedly grow and
   change over time and should be
   reissued several times to a wide
   distribution.

2. Inventory methods must be determined
   and instructions must be written
   covering counting and count recording
   procedures, unit-of-measure problems
   and their resolution, location numbering
   methods, item and lot number
   identification techniques, and methods
   for identifying and counting company
   owned off-site material.  Off-site

material includes material in-transit as well as material temporarily off-site for other reasons.

3. Data processing requirements must be defined including both software specifications and the computing resources and support that will be required during and after the physical counting.

4. Equipment and labor must be scheduled. Equipment requirements may include additional scales, forklifts, and computer terminals that must be purchased or leased. Personnel must be scheduled for counting, writing, auditing, and reconciliation.

5. Cutoff dates and times must be established for receipts, issues, shop movements, shipments to customers, shipments to other plants and warehouses, and scrap processing.

6. A fresh standard cost buildup should be done shortly before the inventory takes place. This requires verification of labor rates and a freeze on routing and vendor price changes.

7. Shortly before the physical inventory begins, counting responsibilities must be defined and supervisors and auditors must be trained. On the day before the count begins, or on the first morning of the count itself, counters and writers must be trained and given an opportunity to ask questions. The value of training cannot be overemphasized. Some companies even set up miniature stockrooms in the training area to allow hands-on demonstration of the required methods and procedures.

## *Physical Inventory Methods*

Many businesses, even those that control material movements and cycle counts with on-line radio-based system, run off-line physical inventories. There are often many more counting teams than radios, and the cost of purchasing additional radios is simply not justified for an activity that takes place infrequently. Further, there is little need for on-line database updating during a physical inventory.

Physical inventory taking methods differ significantly from one business to another based on the products and facilities involved and the requirements of both management

```
7/14            PHYSICAL INVENTORY COUNT SHEET      PAGE 6 OF 32

Location                        Item          Quantity    Lot No.

A-14-D-01
||||||||||||||||||||||||        _____   _____   _____
                                _____   _____   _____
                                _____   _____   _____
                                _____   _____   _____

A-14-D-02
||||||||||||||||||||||||        _____   _____   _____
                                _____   _____   _____
                                _____   _____   _____
                                _____   _____   _____

A-14-D-03
||||||||||||||||||||||||        _____   _____   _____
                                _____   _____   _____
                                _____   _____   _____
                                _____   _____   _____

A-14-D-04
||||||||||||||||||||||||        _____   _____   _____
                                _____   _____   _____
                                _____   _____   _____
                                _____   _____   _____

A-14-D-05
||||||||||||||||||||||||        _____   _____   _____
                                _____   _____   _____
                                _____   _____   _____
                                _____   _____   _____

A-14D-06
||||||||||||||||||||||||        _____   _____   _____
                                _____   _____   _____
                                _____   _____   _____
                                _____   _____   _____

A-14-D-07
||||||||||||||||||||||||        _____   _____   _____
                                _____   _____   _____
                                _____   _____   _____
                                _____   _____   _____
```

and the auditors. The sample method
described below will support high levels of
accuracy in the completed inventory but may
not meet the needs of all companies.

Assuming that radio terminals are not to be
used, a physical inventory system can be

based on clerks working at on-line CRT workstations with attached bar code wands. The clerks are responsible for distributing the counting work to two-person counting teams. The counting process begins after all material movement in the facility has ceased. Clerks scan the bar-coded employee badges of the counting team members, one team at a time. The inventory system then prints count sheets like the sample on the previous page for each team. It also records the identity of the people who are to count each location for future reference.

Count sheets are usually constructed by the inventory control system to provide approximately 2 hours of work for a team. Factors involved in estimating the workload for a group of inventory locations can include the accessibility of locations, the nature and quantity of items stored in the locations, and the necessity for special equipment such as lift trucks or scales. Each count sheet can consist of several pages.

Along with the count sheets, the system prints a group of simple count identification labels as shown at the right. These labels simplify some aspects

| LOCATION |
| --- |
| A-14-D-01 |
| COUNTED 7/14 BY |
| J ADAMS |
| R SMITH |

of the auditing function and also provide a measure of motivation to counters by labeling each location with their names. The count sheets and the count identification labels are given to the team. The team proceeds into the warehouse and begins counting.

The actual counting process is done one location at a time, with each team responsible for the identification and counting of all the material in a location. As items are identified and counted, one member of the team records item numbers and quantities on the count sheet, while the other does the physical counting. Upon completion of each location, a count identification label is applied to the material. When all assigned locations have been counted, the team returns to the office and turns in the sheet.

In the office, a clerk rescans the team's employee badges to produce another count sheet for them. Then the clerk keys the information on the completed count sheet. Bar codes are provided to assure accurate entry of location numbers. The system verifies the correctness of the item number, quantity, and lot number information as it is keyed and gives the clerk an opportunity to correct keying errors as variances are found. When a variance is found that is not due to a keying error, the location is added to a file of pending

recounts. When the results of the inventory
match the system's records exactly, the
system logs the count in its history file,
updates the date of last count on the location
file, and continues to the next location.

The counting teams do not use bar codes
because the objective is to collect item
numbers, quantities, and lot numbers without
reference to existing records. The absence of
bar codes on the floor, however, does not
mean that the principle of data validation has
been abandoned. Validation, in this
procedure, comes from the comparison of the
physical count with the system's records. The
location is bar coded on the count sheet not
only to improve key entry productivity but also
because the location number itself isn't
otherwise validated and must be entered
accurately.

After the last count sheet has been printed
and issued to a counting team, the inventory
control system begins printing recounts. Recounting procedures are identical to the initial
counts, except for a restriction that the
recounting team cannot be the team that
performed the initial count. If a recount does
not match the recorded on-hand value but
does match the first count, an inventory
variance is recorded for later posting to the
inventory file. If the recount matches neither

the recorded on-hand balance nor the original
count, the location is added to a problem list.

As the counting and recounting processes
near completion, the auditing function comes
into play. Both the accumulated variances
and the problem list can be displayed on a
terminal and printed in the form of count
sheets as necessary. Auditors and supervisors
can recount locations and enter or adjust the
count results. The system tracks and can
report the number of instances in which a
count made by a counting team is adjusted by
an auditor or by a supervisor.

Eventually it will be decided that the inventory
is adequately complete to allow the facility to
resume work. At this point, the system will
post the accumulated variances to the
inventory records and will place holds on the
locations that remain on the problem list.
These holds can be released one at a time or
in groups by auditors or supervisors as the
problems are resolved.

Following completion of the physical inventory
and any final adjustments, the inventory
control system reports:

➢ Inventory value

➢ Value of adjustments made during the
inventory

➢ Recount listings by part number,
location, and team

➢ Inventory accuracy as measured by the
physical inventory.

This procedure assumes that the count teams
are required to verify item numbers, quantities
and, where they exist, lot numbers.  In some
businesses, item identification is more difficult
and therefore more expensive than in others.
If it is unreasonable to expect a counting team
to determine item numbers by looking at the
product, count sheets can include item
number and descriptions.  However, the teams
should be specifically instructed to verify part
numbers in addition to counting and
recording quantities.

# Conclusions

C ycle counting should not be a major
headache. Certainly some companies
have gone overboard, while others have
avoided cycle counting entirely out of
unnecessary concern for the size of the
commitment required. A realistic cycle
counting program can be modest, and yet be
effective. And its cost can easily be recouped
when the physical inventory is eliminated.

Cycle counting's primary objective is to
measure the level of inventory accuracy and to
give management warning when accuracy
levels start to slip (or climb). A valid secondary
objective is to audit procedures and methods
to assure that everyone is doing their job
correctly. The idea of correcting existing errors
as part of a cycle counting program is valid,
but the overall direct impact of a reasonable
cycle counting program on inventory accuracy
will be small.

The correct amount of cycle counting can be
calculated when the desired level of precision
and needed frequency of measurement is
determined.

Physical Inventories – in which everything in
the warehouse is counted all at once – are

generally obsolete and are necessary only in instances in which the perpetual inventory records are beyond redemption. However, the concepts of physical inventory taking should not be forgotten because there are rare instances in which physical inventories are necessary. Examples include:

- ➢ Physical takeover of a warehouse from a party whose inventory records cannot be trusted

- ➢ Catastrophic loss of inventory records

- ➢ Natural disasters such as fire and flood involving significant inventory loss

If a physical inventory is needed, it is, of course, much better to do it well than poorly.

# Appendix – Tables

| Table 1 | | | | | | | | | |
|---|---|---|---|---|---|---|---|---|---|
| Number of Counts Required to Measure Inventory Accuracy Within a Specified Tolerance | | | | | | | | | |
| Expected Accuracy | Tolerance | | | | | | | | |
| | 1% | 2% | 3% | 4% | 5% | 6% | 7% | 8% | 9% | 10% |
| 99.9% | 90 | 22 | 10 | 6 | 4 | 2 | 2 | 1 | 1 | 1 |
| 99.8% | 180 | 45 | 20 | 11 | 7 | 5 | 4 | 3 | 2 | 2 |
| 99.7% | 269 | 67 | 30 | 17 | 11 | 7 | 5 | 4 | 3 | 3 |
| 99.6% | 359 | 90 | 40 | 22 | 14 | 10 | 7 | 6 | 4 | 4 |
| 99.5% | 448 | 112 | 50 | 28 | 18 | 12 | 9 | 7 | 6 | 4 |
| 99.4% | 537 | 134 | 60 | 34 | 21 | 15 | 11 | 8 | 7 | 5 |
| 99.3% | 626 | 156 | 70 | 39 | 25 | 17 | 13 | 10 | 8 | 6 |
| 99.2% | 714 | 179 | 79 | 45 | 29 | 20 | 15 | 11 | 9 | 7 |
| 99.1% | 803 | 201 | 89 | 50 | 32 | 22 | 16 | 13 | 10 | 8 |
| 99.0% | 891 | 223 | 99 | 56 | 36 | 25 | 18 | 14 | 11 | 9 |
| 98.5% | 1330 | 332 | 148 | 83 | 53 | 37 | 27 | 21 | 16 | 13 |
| 98.0% | 1764 | 441 | 196 | 110 | 71 | 49 | 36 | 28 | 22 | 18 |
| 97.5% | 2194 | 548 | 244 | 137 | 88 | 61 | 45 | 34 | 27 | 22 |
| 97.0% | 2619 | 655 | 291 | 164 | 105 | 73 | 53 | 41 | 32 | 26 |
| 96.5% | 3040 | 760 | 338 | 190 | 122 | 84 | 62 | 47 | 38 | 30 |
| 96.0% | 3456 | 864 | 384 | 216 | 138 | 96 | 71 | 54 | 43 | 35 |
| 95.5% | 3868 | 967 | 430 | 242 | 155 | 107 | 79 | 60 | 48 | 39 |
| 95.0% | 4275 | 1069 | 475 | 267 | 171 | 119 | 87 | 67 | 53 | 43 |
| 94.0% | 5076 | 1269 | 564 | 317 | 203 | 141 | 104 | 79 | 63 | 51 |
| 93.0% | 5859 | 1465 | 651 | 366 | 234 | 163 | 120 | 92 | 72 | 59 |
| 92.0% | 6624 | 1656 | 736 | 414 | 265 | 184 | 135 | 104 | 82 | 66 |
| 91.0% | 7371 | 1843 | 819 | 461 | 295 | 205 | 150 | 115 | 91 | 74 |
| 90.0% | 8100 | 2025 | 900 | 506 | 324 | 225 | 165 | 127 | 100 | 81 |
| 85.0% | 11475 | 2869 | 1275 | 717 | 459 | 319 | 234 | 179 | 142 | 115 |
| 80.0% | 14400 | 3600 | 1600 | 900 | 576 | 400 | 294 | 225 | 178 | 144 |
| 75.0% | 16875 | 4219 | 1875 | 1055 | 675 | 469 | 344 | 264 | 208 | 169 |
| 70.0% | 18900 | 4725 | 2100 | 1181 | 756 | 525 | 386 | 295 | 233 | 189 |
| 65.0% | 20475 | 5119 | 2275 | 1280 | 819 | 569 | 418 | 320 | 253 | 205 |
| 60.0% | 21600 | 5400 | 2400 | 1350 | 864 | 600 | 441 | 338 | 267 | 216 |
| 55.0% | 22275 | 5569 | 2475 | 1392 | 891 | 619 | 455 | 348 | 275 | 223 |
| 50.0% | 22500 | 5625 | 2500 | 1406 | 900 | 625 | 459 | 352 | 278 | 225 |

| Table 2 | | | | | | | | | |
|---|---|---|---|---|---|---|---|---|---|
| Control Limit Spacing (Excluding Confidence Factor) | | | | | | | | | |
| Expected Accuracy | Sample Size | | | | | | | | |
| | 10 | 15 | 20 | 25 | 30 | 35 | 40 | 45 | 50 |
| 99.9% | 1.0 | 0.8 | 0.7 | 0.6 | 0.6 | 0.5 | 0.5 | 0.5 | 0.4 |
| 99.8% | 1.4 | 1.2 | 1.0 | 0.9 | 0.8 | 0.8 | 0.7 | 0.7 | 0.6 |
| 99.7% | 1.7 | 1.4 | 1.2 | 1.1 | 1.0 | 0.9 | 0.9 | 0.8 | 0.8 |
| 99.6% | 2.0 | 1.6 | 1.4 | 1.3 | 1.2 | 1.1 | 1.0 | 0.9 | 0.9 |
| 99.5% | 2.2 | 1.8 | 1.6 | 1.4 | 1.3 | 1.2 | 1.1 | 1.1 | 1.0 |
| 99.4% | 2.4 | 2.0 | 1.7 | 1.5 | 1.4 | 1.3 | 1.2 | 1.2 | 1.1 |
| 99.3% | 2.6 | 2.2 | 1.9 | 1.7 | 1.5 | 1.4 | 1.3 | 1.2 | 1.2 |
| 99.2% | 2.8 | 2.3 | 2.0 | 1.8 | 1.6 | 1.5 | 1.4 | 1.3 | 1.3 |
| 99.1% | 3.0 | 2.4 | 2.1 | 1.9 | 1.7 | 1.6 | 1.5 | 1.4 | 1.3 |
| 99.0% | 3.1 | 2.6 | 2.2 | 2.0 | 1.8 | 1.7 | 1.6 | 1.5 | 1.4 |
| 98.5% | 3.8 | 3.1 | 2.7 | 2.4 | 2.2 | 2.1 | 1.9 | 1.8 | 1.7 |
| 98.0% | 4.4 | 3.6 | 3.1 | 2.8 | 2.6 | 2.4 | 2.2 | 2.1 | 2.0 |
| 97.5% | 4.9 | 4.0 | 3.5 | 3.1 | 2.9 | 2.6 | 2.5 | 2.3 | 2.2 |
| 97.0% | 5.4 | 4.4 | 3.8 | 3.4 | 3.1 | 2.9 | 2.7 | 2.5 | 2.4 |
| 96.5% | 5.8 | 4.7 | 4.1 | 3.7 | 3.4 | 3.1 | 2.9 | 2.7 | 2.6 |
| 96.0% | 6.2 | 5.1 | 4.4 | 3.9 | 3.6 | 3.3 | 3.1 | 2.9 | 2.8 |
| 95.5% | 6.6 | 5.4 | 4.6 | 4.1 | 3.8 | 3.5 | 3.3 | 3.1 | 2.9 |
| 95.0% | 6.9 | 5.6 | 4.9 | 4.4 | 4.0 | 3.7 | 3.4 | 3.2 | 3.1 |
| 94.0% | 7.5 | 6.1 | 5.3 | 4.7 | 4.3 | 4.0 | 3.8 | 3.5 | 3.4 |
| 93.0% | 8.1 | 6.6 | 5.7 | 5.1 | 4.7 | 4.3 | 4.0 | 3.8 | 3.6 |
| 92.0% | 8.6 | 7.0 | 6.1 | 5.4 | 5.0 | 4.6 | 4.3 | 4.0 | 3.8 |
| 91.0% | 9.0 | 7.4 | 6.4 | 5.7 | 5.2 | 4.8 | 4.5 | 4.3 | 4.0 |
| 90.0% | 9.5 | 7.7 | 6.7 | 6.0 | 5.5 | 5.1 | 4.7 | 4.5 | 4.2 |
| 85.0% | 11.3 | 9.2 | 8.0 | 7.1 | 6.5 | 6.0 | 5.6 | 5.3 | 5.0 |
| 80.0% | 12.6 | 10.3 | 8.9 | 8.0 | 7.3 | 6.8 | 6.3 | 6.0 | 5.7 |
| 75.0% | 13.7 | 11.2 | 9.7 | 8.7 | 7.9 | 7.3 | 6.8 | 6.5 | 6.1 |
| 70.0% | 14.5 | 11.8 | 10.2 | 9.2 | 8.4 | 7.7 | 7.2 | 6.8 | 6.5 |
| 65.0% | 15.1 | 12.3 | 10.7 | 9.5 | 8.7 | 8.1 | 7.5 | 7.1 | 6.7 |
| 60.0% | 15.5 | 12.6 | 11.0 | 9.8 | 8.9 | 8.3 | 7.7 | 7.3 | 6.9 |
| 55.0% | 15.7 | 12.8 | 11.1 | 9.9 | 9.1 | 8.4 | 7.9 | 7.4 | 7.0 |
| 50.0% | 15.8 | 12.9 | 11.2 | 10.0 | 9.1 | 8.5 | 7.9 | 7.5 | 7.1 |

Table Continued on Page 69

| Table 2 (Continued) | | | | | | | | | |
|---|---|---|---|---|---|---|---|---|---|
| Control Limit Spacing (Excluding Confidence Factor) | | | | | | | | | |
| Expected Accuracy | Sample Size | | | | | | | | |
| | 60 | 70 | 80 | 90 | 100 | 110 | 120 | 130 | 140 | 150 |
| 99.9% | 0.4 | 0.4 | 0.4 | 0.3 | 0.3 | 0.3 | 0.3 | 0.3 | 0.3 | 0.3 |
| 99.8% | 0.6 | 0.5 | 0.5 | 0.5 | 0.4 | 0.4 | 0.4 | 0.4 | 0.4 | 0.4 |
| 99.7% | 0.7 | 0.7 | 0.6 | 0.6 | 0.5 | 0.5 | 0.5 | 0.5 | 0.5 | 0.4 |
| 99.6% | 0.8 | 0.8 | 0.7 | 0.7 | 0.6 | 0.6 | 0.6 | 0.6 | 0.5 | 0.5 |
| 99.5% | 0.9 | 0.8 | 0.8 | 0.7 | 0.7 | 0.7 | 0.6 | 0.6 | 0.6 | 0.6 |
| 99.4% | 1.0 | 0.9 | 0.9 | 0.8 | 0.8 | 0.7 | 0.7 | 0.7 | 0.7 | 0.6 |
| 99.3% | 1.1 | 1.0 | 0.9 | 0.9 | 0.8 | 0.8 | 0.8 | 0.7 | 0.7 | 0.7 |
| 99.2% | 1.2 | 1.1 | 1.0 | 0.9 | 0.9 | 0.8 | 0.8 | 0.8 | 0.8 | 0.7 |
| 99.1% | 1.2 | 1.1 | 1.1 | 1.0 | 0.9 | 0.9 | 0.9 | 0.8 | 0.8 | 0.8 |
| 99.0% | 1.3 | 1.2 | 1.1 | 1.0 | 1.0 | 0.9 | 0.9 | 0.9 | 0.8 | 0.8 |
| 98.5% | 1.6 | 1.5 | 1.4 | 1.3 | 1.2 | 1.2 | 1.1 | 1.1 | 1.0 | 1.0 |
| 98.0% | 1.8 | 1.7 | 1.6 | 1.5 | 1.4 | 1.3 | 1.3 | 1.2 | 1.2 | 1.1 |
| 97.5% | 2.0 | 1.9 | 1.7 | 1.6 | 1.6 | 1.5 | 1.4 | 1.4 | 1.3 | 1.3 |
| 97.0% | 2.2 | 2.0 | 1.9 | 1.8 | 1.7 | 1.6 | 1.6 | 1.5 | 1.4 | 1.4 |
| 96.5% | 2.4 | 2.2 | 2.1 | 1.9 | 1.8 | 1.8 | 1.7 | 1.6 | 1.6 | 1.5 |
| 96.0% | 2.5 | 2.3 | 2.2 | 2.1 | 2.0 | 1.9 | 1.8 | 1.7 | 1.7 | 1.6 |
| 95.5% | 2.7 | 2.5 | 2.3 | 2.2 | 2.1 | 2.0 | 1.9 | 1.8 | 1.8 | 1.7 |
| 95.0% | 2.8 | 2.6 | 2.4 | 2.3 | 2.2 | 2.1 | 2.0 | 1.9 | 1.8 | 1.8 |
| 94.0% | 3.1 | 2.8 | 2.7 | 2.5 | 2.4 | 2.3 | 2.2 | 2.1 | 2.0 | 1.9 |
| 93.0% | 3.3 | 3.0 | 2.9 | 2.7 | 2.6 | 2.4 | 2.3 | 2.2 | 2.2 | 2.1 |
| 92.0% | 3.5 | 3.2 | 3.0 | 2.9 | 2.7 | 2.6 | 2.5 | 2.4 | 2.3 | 2.2 |
| 91.0% | 3.7 | 3.4 | 3.2 | 3.0 | 2.9 | 2.7 | 2.6 | 2.5 | 2.4 | 2.3 |
| 90.0% | 3.9 | 3.6 | 3.4 | 3.2 | 3.0 | 2.9 | 2.7 | 2.6 | 2.5 | 2.4 |
| 85.0% | 4.6 | 4.3 | 4.0 | 3.8 | 3.6 | 3.4 | 3.3 | 3.1 | 3.0 | 2.9 |
| 80.0% | 5.2 | 4.8 | 4.5 | 4.2 | 4.0 | 3.8 | 3.7 | 3.5 | 3.4 | 3.3 |
| 75.0% | 5.6 | 5.2 | 4.8 | 4.6 | 4.3 | 4.1 | 4.0 | 3.8 | 3.7 | 3.5 |
| 70.0% | 5.9 | 5.5 | 5.1 | 4.8 | 4.6 | 4.4 | 4.2 | 4.0 | 3.9 | 3.7 |
| 65.0% | 6.2 | 5.7 | 5.3 | 5.0 | 4.8 | 4.5 | 4.4 | 4.2 | 4.0 | 3.9 |
| 60.0% | 6.3 | 5.9 | 5.5 | 5.2 | 4.9 | 4.7 | 4.5 | 4.3 | 4.1 | 4.0 |
| 55.0% | 6.4 | 5.9 | 5.6 | 5.2 | 5.0 | 4.7 | 4.5 | 4.4 | 4.2 | 4.1 |
| 50.0% | 6.5 | 6.0 | 5.6 | 5.3 | 5.0 | 4.8 | 4.6 | 4.4 | 4.2 | 4.1 |
| Table Continued on Page 70 | | | | | | | | | |

| Table 2 (Continued) | | | | | | | | | |
|---|---|---|---|---|---|---|---|---|---|
| Control Limit Spacing (Excluding Confidence Factor) | | | | | | | | | |
| Expected Accuracy | Sample Size | | | | | | | | |
| | 200 | 250 | 300 | 350 | 400 | 450 | 500 | 550 | 600 | 650 |
| 99.9% | 0.2 | 0.2 | 0.2 | 0.2 | 0.2 | 0.1 | 0.1 | 0.1 | 0.1 | 0.1 |
| 99.8% | 0.3 | 0.3 | 0.3 | 0.2 | 0.2 | 0.2 | 0.2 | 0.2 | 0.2 | 0.2 |
| 99.7% | 0.4 | 0.3 | 0.3 | 0.3 | 0.3 | 0.3 | 0.2 | 0.2 | 0.2 | 0.2 |
| 99.6% | 0.4 | 0.4 | 0.4 | 0.3 | 0.3 | 0.3 | 0.3 | 0.3 | 0.3 | 0.2 |
| 99.5% | 0.5 | 0.4 | 0.4 | 0.4 | 0.4 | 0.3 | 0.3 | 0.3 | 0.3 | 0.3 |
| 99.4% | 0.5 | 0.5 | 0.4 | 0.4 | 0.4 | 0.4 | 0.3 | 0.3 | 0.3 | 0.3 |
| 99.3% | 0.6 | 0.5 | 0.5 | 0.4 | 0.4 | 0.4 | 0.4 | 0.4 | 0.3 | 0.3 |
| 99.2% | 0.6 | 0.6 | 0.5 | 0.5 | 0.4 | 0.4 | 0.4 | 0.4 | 0.4 | 0.3 |
| 99.1% | 0.7 | 0.6 | 0.5 | 0.5 | 0.5 | 0.4 | 0.4 | 0.4 | 0.4 | 0.4 |
| 99.0% | 0.7 | 0.6 | 0.6 | 0.5 | 0.5 | 0.5 | 0.4 | 0.4 | 0.4 | 0.4 |
| 98.5% | 0.9 | 0.8 | 0.7 | 0.6 | 0.6 | 0.6 | 0.5 | 0.5 | 0.5 | 0.5 |
| 98.0% | 1.0 | 0.9 | 0.8 | 0.7 | 0.7 | 0.7 | 0.6 | 0.6 | 0.6 | 0.5 |
| 97.5% | 1.1 | 1.0 | 0.9 | 0.8 | 0.8 | 0.7 | 0.7 | 0.7 | 0.6 | 0.6 |
| 97.0% | 1.2 | 1.1 | 1.0 | 0.9 | 0.9 | 0.8 | 0.8 | 0.7 | 0.7 | 0.7 |
| 96.5% | 1.3 | 1.2 | 1.1 | 1.0 | 0.9 | 0.9 | 0.8 | 0.8 | 0.8 | 0.7 |
| 96.0% | 1.4 | 1.2 | 1.1 | 1.0 | 1.0 | 0.9 | 0.9 | 0.8 | 0.8 | 0.8 |
| 95.5% | 1.5 | 1.3 | 1.2 | 1.1 | 1.0 | 1.0 | 0.9 | 0.9 | 0.8 | 0.8 |
| 95.0% | 1.5 | 1.4 | 1.3 | 1.2 | 1.1 | 1.0 | 1.0 | 0.9 | 0.9 | 0.9 |
| 94.0% | 1.7 | 1.5 | 1.4 | 1.3 | 1.2 | 1.1 | 1.1 | 1.0 | 1.0 | 0.9 |
| 93.0% | 1.8 | 1.6 | 1.5 | 1.4 | 1.3 | 1.2 | 1.1 | 1.1 | 1.0 | 1.0 |
| 92.0% | 1.9 | 1.7 | 1.6 | 1.5 | 1.4 | 1.3 | 1.2 | 1.2 | 1.1 | 1.1 |
| 91.0% | 2.0 | 1.8 | 1.7 | 1.5 | 1.4 | 1.3 | 1.3 | 1.2 | 1.2 | 1.1 |
| 90.0% | 2.1 | 1.9 | 1.7 | 1.6 | 1.5 | 1.4 | 1.3 | 1.3 | 1.2 | 1.2 |
| 85.0% | 2.5 | 2.3 | 2.1 | 1.9 | 1.8 | 1.7 | 1.6 | 1.5 | 1.5 | 1.4 |
| 80.0% | 2.8 | 2.5 | 2.3 | 2.1 | 2.0 | 1.9 | 1.8 | 1.7 | 1.6 | 1.6 |
| 75.0% | 3.1 | 2.7 | 2.5 | 2.3 | 2.2 | 2.0 | 1.9 | 1.8 | 1.8 | 1.7 |
| 70.0% | 3.2 | 2.9 | 2.6 | 2.4 | 2.3 | 2.2 | 2.0 | 2.0 | 1.9 | 1.8 |
| 65.0% | 3.4 | 3.0 | 2.8 | 2.5 | 2.4 | 2.2 | 2.1 | 2.0 | 1.9 | 1.9 |
| 60.0% | 3.5 | 3.1 | 2.8 | 2.6 | 2.4 | 2.3 | 2.2 | 2.1 | 2.0 | 1.9 |
| 55.0% | 3.5 | 3.1 | 2.9 | 2.7 | 2.5 | 2.3 | 2.2 | 2.1 | 2.0 | 2.0 |
| 50.0% | 3.5 | 3.2 | 2.9 | 2.7 | 2.5 | 2.4 | 2.2 | 2.1 | 2.0 | 2.0 |
| Table Continued on Page 71 | | | | | | | | | |

Table Continued on Page 71

| Table 2 (Continued) | | | | | | | | | |
|---|---|---|---|---|---|---|---|---|---|
| Control Limit Spacing (Excluding Confidence Factor) | | | | | | | | | |
| Expected Accuracy | Sample Size | | | | | | | | |
| | 750 | 1000 | 1250 | 1500 | 1750 | 2000 | 3000 | 5000 | 8000 | 10000 |
| 99.9% | 0.1 | 0.1 | 0.1 | 0.1 | 0.1 | 0.1 | 0.1 | 0.0 | 0.0 | 0.0 |
| 99.8% | 0.2 | 0.1 | 0.1 | 0.1 | 0.1 | 0.1 | 0.1 | 0.1 | 0.0 | 0.0 |
| 99.7% | 0.2 | 0.2 | 0.2 | 0.1 | 0.1 | 0.1 | 0.1 | 0.1 | 0.1 | 0.1 |
| 99.6% | 0.2 | 0.2 | 0.2 | 0.2 | 0.2 | 0.1 | 0.1 | 0.1 | 0.1 | 0.1 |
| 99.5% | 0.3 | 0.2 | 0.2 | 0.2 | 0.2 | 0.2 | 0.1 | 0.1 | 0.1 | 0.1 |
| 99.4% | 0.3 | 0.2 | 0.2 | 0.2 | 0.2 | 0.2 | 0.1 | 0.1 | 0.1 | 0.1 |
| 99.3% | 0.3 | 0.3 | 0.2 | 0.2 | 0.2 | 0.2 | 0.2 | 0.1 | 0.1 | 0.1 |
| 99.2% | 0.3 | 0.3 | 0.3 | 0.2 | 0.2 | 0.2 | 0.2 | 0.1 | 0.1 | 0.1 |
| 99.1% | 0.3 | 0.3 | 0.3 | 0.2 | 0.2 | 0.2 | 0.2 | 0.1 | 0.1 | 0.1 |
| 99.0% | 0.4 | 0.3 | 0.3 | 0.3 | 0.2 | 0.2 | 0.2 | 0.1 | 0.1 | 0.1 |
| 98.5% | 0.4 | 0.4 | 0.3 | 0.3 | 0.3 | 0.3 | 0.2 | 0.2 | 0.1 | 0.1 |
| 98.0% | 0.5 | 0.4 | 0.4 | 0.4 | 0.3 | 0.3 | 0.3 | 0.2 | 0.2 | 0.1 |
| 97.5% | 0.6 | 0.5 | 0.4 | 0.4 | 0.4 | 0.3 | 0.3 | 0.2 | 0.2 | 0.2 |
| 97.0% | 0.6 | 0.5 | 0.5 | 0.4 | 0.4 | 0.4 | 0.3 | 0.2 | 0.2 | 0.2 |
| 96.5% | 0.7 | 0.6 | 0.5 | 0.5 | 0.4 | 0.4 | 0.3 | 0.3 | 0.2 | 0.2 |
| 96.0% | 0.7 | 0.6 | 0.6 | 0.5 | 0.5 | 0.4 | 0.4 | 0.3 | 0.2 | 0.2 |
| 95.5% | 0.8 | 0.7 | 0.6 | 0.5 | 0.5 | 0.5 | 0.4 | 0.3 | 0.2 | 0.2 |
| 95.0% | 0.8 | 0.7 | 0.6 | 0.6 | 0.5 | 0.5 | 0.4 | 0.3 | 0.2 | 0.2 |
| 94.0% | 0.9 | 0.8 | 0.7 | 0.6 | 0.6 | 0.5 | 0.4 | 0.3 | 0.3 | 0.2 |
| 93.0% | 0.9 | 0.8 | 0.7 | 0.7 | 0.6 | 0.6 | 0.5 | 0.4 | 0.3 | 0.3 |
| 92.0% | 1.0 | 0.9 | 0.8 | 0.7 | 0.6 | 0.6 | 0.5 | 0.4 | 0.3 | 0.3 |
| 91.0% | 1.0 | 0.9 | 0.8 | 0.7 | 0.7 | 0.6 | 0.5 | 0.4 | 0.3 | 0.3 |
| 90.0% | 1.1 | 0.9 | 0.8 | 0.8 | 0.7 | 0.7 | 0.5 | 0.4 | 0.3 | 0.3 |
| 85.0% | 1.3 | 1.1 | 1.0 | 0.9 | 0.9 | 0.8 | 0.7 | 0.5 | 0.4 | 0.4 |
| 80.0% | 1.5 | 1.3 | 1.1 | 1.0 | 1.0 | 0.9 | 0.7 | 0.6 | 0.4 | 0.4 |
| 75.0% | 1.6 | 1.4 | 1.2 | 1.1 | 1.0 | 1.0 | 0.8 | 0.6 | 0.5 | 0.4 |
| 70.0% | 1.7 | 1.4 | 1.3 | 1.2 | 1.1 | 1.0 | 0.8 | 0.6 | 0.5 | 0.5 |
| 65.0% | 1.7 | 1.5 | 1.3 | 1.2 | 1.1 | 1.1 | 0.9 | 0.7 | 0.5 | 0.5 |
| 60.0% | 1.8 | 1.5 | 1.4 | 1.3 | 1.2 | 1.1 | 0.9 | 0.7 | 0.5 | 0.5 |
| 55.0% | 1.8 | 1.6 | 1.4 | 1.3 | 1.2 | 1.1 | 0.9 | 0.7 | 0.6 | 0.5 |
| 50.0% | 1.8 | 1.6 | 1.4 | 1.3 | 1.2 | 1.1 | 0.9 | 0.7 | 0.6 | 0.5 |

| Table 3 | |
| :-- | :-- |
| Control Limit Confidence Values | |
| Factor Value | Confidence that a point below the control limit is not caused by random variation (%) |
| 0.0 | 50.0 |
| 0.1 | 54.0 |
| 0.2 | 57.9 |
| 0.3 | 61.8 |
| 0.4 | 65.5 |
| 0.5 | 69.1 |
| 0.6 | 72.6 |
| 0.7 | 75.8 |
| 0.8 | 78.8 |
| 0.9 | 81.6 |
| 1.0 | 84.1 |
| 1.1 | 86.4 |
| 1.2 | 88.5 |
| 1.3 | 90.3 |
| 1.4 | 91.9 |
| 1.5 | 93.3 |
| 1.6 | 94.5 |
| 1.7 | 95.5 |
| 1.8 | 96.4 |
| 1.9 | 97.1 |
| 2.0 | 97.7 |
| 2.1 | 98.2 |
| 2.2 | 98.6 |
| 2.3 | 98.9 |
| 2.4 | 99.2 |
| 2.5 | 99.4 |
| 2.6 | 99.5 |
| 2.7 | 99.7 |
| 2.8 | 99.7 |
| 2.9 | 99.8 |
| 3.0 | 99.9 |

40987753R00046

Made in the USA
Lexington, KY
24 April 2015